Florian Ion **Petrescu &** *Relly Victoria* **Petrescu**

GREEN ENERGY

Germany 2012

Scientific reviewer:

Dr. Veturia CHIROIU
Honorific member of
Technical Sciences Academy of Romania (ASTR)
PhD supervisor in Mechanical Engineering

Copyright

Title book: Green Energy

Author book: Florian Ion Petrescu, Relly Victoria Petrescu

© **2011-2012, Florian Ion PETRESCU**

petrescuflorian@yahoo.com

ALL RIGHTS RESERVED. This book contains material protected under International and Federal Copyright Laws and Treaties. Any unauthorized reprint or use of this material is prohibited. No part of this book may be reproduced or transmitted in any form or by any means, electronic or mechanical, including photocopying, recording, or by any information storage and retrieval system without express written permission from the authors / publisher.

**Manufactured and published by:
Books on Demand GmbH, Norderstedt
ISBN 978-3-8482-2363-3**

Welcome! A Short Book Description:

This book, try a vision about the tomorrow Earth's energies.

Besides the traditional energies, are presented and possible new energies source.

The author presents shortly, the annihilation process, which can donate energies obtained by the annihilation process between a particle and its antiparticle.

Practically it proposes obtaining energy by the process of annihilation of matter with the antimatter.

Another proposed method is the acquiring energy from the source and retransmitting it on the Earth in concentrated form.

Introduction

Energy development is the effort to provide sufficient primary energy sources and secondary energy forms for supply, cost, impact on air pollution and water pollution, mitigation of climate change with renewable energy.

Technologically advanced societies have become increasingly dependent on external energy sources for transportation, the production of many manufactured goods, and the delivery of energy services.

This energy allows people who can afford the cost to live under otherwise unfavorable climatic conditions through the use of heating, ventilation, and/or air conditioning. Level of use of external energy sources differs across societies, as do the climate, convenience, levels of traffic congestion, pollution and availability of domestic energy sources.

All terrestrial energy sources except nuclear, geothermal and tidal are from current solar insolation or from fossil remains of plant and animal life that relied directly and indirectly upon sunlight, respectively.

Ultimately, solar energy itself is the result of the Sun's nuclear fusion.

Geothermal power from hot, hardened rock above the magma of the Earth's core is the result of the decay of radioactive materials present beneath the Earth's crust, and nuclear fission relies on man-made fission of heavy radioactive elements in the Earth's crust; in both cases these elements were produced in supernova explosions before the formation of the solar system.

Renewable energy is energy which comes from natural resources such as sunlight, wind, rain, tides, and geothermal heat, which are renewable (naturally replenished).

In 2008, about 19% of global final energy consumption came from renewables, with 13% coming from traditional biomass, which is mainly used for heating, and 3.2% from hydroelectricity.

New renewables (small hydro, modern biomass, wind, solar, geothermal, and biofuels) accounted for another 2.7% and are growing very rapidly.

The share of renewables in electricity generation is around 18%, with 15% of global electricity coming from hydroelectricity and 3% from new renewables. Wind power is growing at the rate of 30% annually, with a worldwide installed capacity of 158 (GW) in 2009, and is widely used in Europe, Asia, and the United States.

At the end of 2009, cumulative global photovoltaic (PV) installations surpassed 21 GW and PV power stations are popular in Germany and Spain.

Solar thermal power stations operate in the USA and Spain, and the largest of these is the 354 megawatt (MW) SEGS power plant in the Mojave Desert.

The world's largest geothermal power installation is The Geysers in California, with a rated capacity of 750 MW. Brazil has one of the largest renewable energy programs in the world, involving production of ethanol fuel from sugar cane, and ethanol now provides 18% of the country's automotive fuel.

Ethanol fuel is also widely available in the USA, the world's largest producer in absolute terms, although not as a percentage of its total motor fuel use.

While many renewable energy projects are large-scale, renewable technologies are also suited to rural and remote areas, where energy is often crucial in human development.

Globally, an estimated 3 million households get power from small solar PV systems. Micro-hydro systems configured into village-scale or county-scale mini-grids serve many areas.

More than 30 million rural households get lighting and cooking from biogas made in household-scale digesters. Biomass cook stoves are used by 160 million households.

Climate change concerns, coupled with high oil prices, peak oil, and increasing government support, are driving increasing renewable energy legislation, incentives and commercialization.

New government spending, regulation and policies helped the industry weather the 2009 economic crisis better than many other sectors.

First energy source

Life's First Energy Source

An obscure compound known as pyrophosphite could have been a source of energy that allowed the first life on Earth to form.

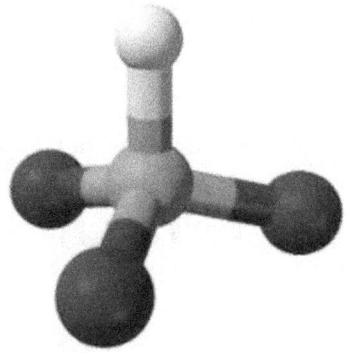

Researchers at the University of Leeds have uncovered new clues to the origins of life on Earth.

The team found that a compound known as pyrophosphite may have been an important energy source for primitive life forms.

There are several conflicting theories of how life on Earth emerged from inanimate matter billions of years ago – a process known as abiogenesis.

"It's a chicken and egg question," said Dr Terry Kee of the University of Leeds, who led the research. "Scientists are in disagreement over what came first – replication, or metabolism. But there is a third part to the equation – and that is energy."

All living things require a continual supply of energy in order to function. This energy is carried around our bodies within certain molecules, one of the best known being ATP*, which converts heat from the sun into a useable form for animals and plants.

At any one time, the human body contains just 250g of ATP – this provides roughly the same amount of energy as a single AA battery. This ATP store is being constantly used and regenerated in cells via a process known as respiration, which is driven by natural catalysts called enzymes.

"You need enzymes to make ATP and you need ATP to make enzymes," explained Dr Kee. "The question is: where did energy come from before either of these two things existed? We think that the answer may lie in simple molecules such as pyrophosphite which is chemically very similar to ATP, but has the potential to transfer energy without enzymes."

The key to the battery-like properties of both ATP and pyrophosphite is an element called phosphorus, which is essential for all living things. Not only is phosphorus the active component of ATP, it also forms the backbone of DNA and is important in the structure of cell walls.

But despite its importance to life, it is not fully understood how phosphorus first appeared in our atmosphere. One theory is that it was contained within the many meteorites that collided with the Earth billions of years ago.

"Phosphorus is present within several meteoritic minerals and it is possible that this reacted to form pyrophosphite under the acidic, volcanic conditions of early Earth," added Dr Kee.

The findings, published in the journal Chemical Communications, are the first to suggest that pyrophosphite may have been relevant in the shift from basic chemistry to complex biology when life on earth began. Since completing this research, Dr Kee and his team have found even further evidence for the importance of this molecule and now hope to team up with collaborators from NASA to investigate its role in abiogenesis.

Human mitochondrial genetics is the study of the genetics of the DNA contained in human mitochondria. Mitochondria are small structures in cells that generate energy for the cell to use, and are hence referred to as the "powerhouses" of the cell.

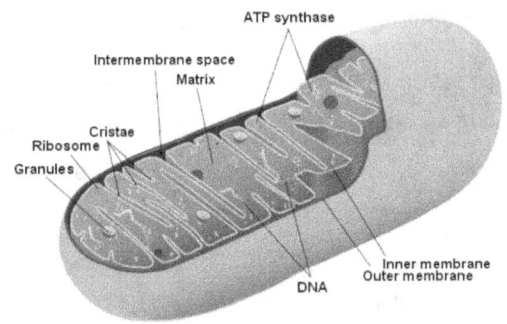

Mitochondrial DNA (mtDNA) is not transmitted through nuclear DNA (nDNA). In humans, as in most multicellular organisms, mitochondrial DNA is inherited only from the mother's ovum.

Mitochondrial inheritance is therefore non-Mendelian, as Mendelian inheritance presumes that half the genetic material of a fertilized egg (zygote) derives from each parent.

Eighty percent of mitochondrial DNA codes for functional mitochondrial proteins, and therefore most mitochondrial DNA mutations lead to functional problems, which may be manifested as muscle disorders (myopathies).

Understanding the genetic mutations that affect mitochondria can help us to understand the inner workings of cells and organisms, as well as helping to suggest methods for successful therapeutic tissue and organ cloning, and to treatments or possibly cures for many devastating muscular disorders.

Because they provide 36 molecules of ATP per glucose molecule in contrast to the 2 ATP molecules produced by glycolysis, mitochondria are essential to all higher organisms for sustaining life. The mitochondrial diseases are genetic disorders carried specifically in mitochondrial DNA; slight problems with any one of the numerous enzymes used by the mitochondria can be devastating to the cell, and in turn, to the organism.

The pyrophosphite and human mitochondria are the principal motors of the human energetic processes.

We should better understand these processes, to can prolong our life.

The oldest energy source

Man started to use biomass for energy on the day that our ancestors discovered fire, and used it for cooking. Biomass is actually just another word for biological-mass. Biomass is anything that has been grown or has lived, except for fossil fuels (coal, oil, natural gas etc). Fossil fuels were of course created by the decay of living organisms many millennia ago in pre-history and are biomass in that sense, but these are not included within the term 'biomass' as used by renewable energy experts.

Biomass takes many forms; some of the most well known are: wood, straw, bio waste, wood chip, waste paper, organic slurries from the processing of foodstuffs, livestock farming, sewage treatment, etc.

So biomass can also be grown as a crop for use as fuel. If the biomass is to be grown it will need to be selected to be of high calorific value (give of lots of heat when burnt), grow fast, need little fertilizing or watering, require low power requirements during growing and be cheaply harvested. However, the growing of biomass to use as biofuel on a large scale would have the effect of reducing available land for food crops.

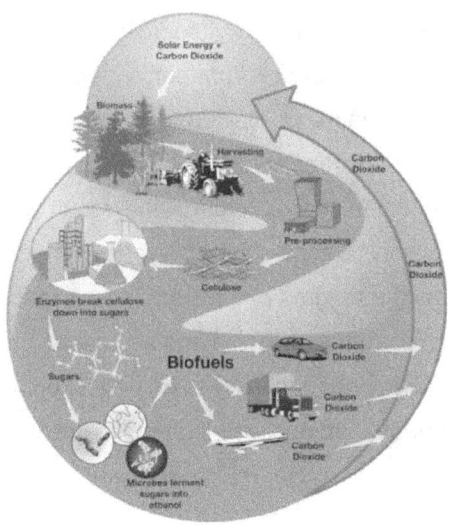

New energy sources

Energy development is the effort to provide sufficient primary energy sources and secondary energy forms for supply, cost, impact on air pollution and water pollution, mitigation of climate change with renewable energy.

Technologically advanced societies have become increasingly dependent on external energy sources for transportation, the production of many manufactured goods, and the delivery of energy services. This energy allows people who can afford the cost to live under otherwise unfavorable climatic conditions through the use of heating, ventilation, and/or air conditioning.

All terrestrial energy sources except nuclear, geothermal and tidal are from current solar insolation or from fossil remains of plant and animal life that relied directly and indirectly upon sunlight, respectively. Ultimately, solar energy itself is the result of the Sun's nuclear fusion. Geothermal power from hot, hardened rock above the magma of the Earth's core is the result of the decay of radioactive materials present beneath the Earth's crust, and nuclear fission relies on man-made fission of heavy radioactive elements in the Earth's crust; in both cases these elements were produced in supernova explosions before the formation of the solar system.

Wind power is growing at the rate of 30% annually, with a worldwide installed capacity of 158 (GW) in 2009, and is widely used in Europe, Asia, and the United States.

At the end of 2009, cumulative global photovoltaic (PV) installations surpassed 21 GW and PV power stations are popular in Germany and Spain. Solar thermal power stations operate in the USA and Spain, and the largest of these is the 354 megawatt (MW) SEGS power plant in the Mojave Desert.

The world's largest geothermal power installation is The Geysers in California, with a rated capacity of 750 MW.

Brazil has one of the largest renewable energy programs in the world, involving production of ethanol fuel from sugar cane, and ethanol now provides 18% of the country's automotive fuel.

Ethanol fuel is also widely available in the USA, the world's largest producer in absolute terms, although not as a percentage of its total motor fuel use.

While many renewable energy projects are large-scale, renewable technologies are also suited to rural and remote areas, where energy is often crucial in human development. Globally, an estimated 3 million households get power from small solar PV systems. Micro-hydro systems configured into village-scale or county-scale mini-grids serve many areas.

More than 30 million rural households get lighting and cooking from biogas made in household-scale digesters. Biomass cook stoves are used by 160 million households.

MAINSTREAM FORMS OF RENEWABLE ENERGY

- 1. Wind power
- 2. Hydropower
- 3. Solar energy
- 4. Biomass
- 5. Biofuel
- 6. Geothermal energy
- 7. Tidal

- 8. Hydrogen obtained by Artificial photosynthesis
- 9. Black light Power
- 10. Waves Power

1. Wind power

Airflows can be used to run wind turbines. Modern wind turbines range from around 600 kW to 5 MW of rated power, although turbines with rated output of 1.5–3 MW have become the most common for commercial use; the power output of a turbine is a function of the cube of the wind speed, so as wind speed increases, power output increases dramatically. Typical capacity factors are 20-40%, with values at the upper end of the range in particularly favorable sites [1].

2. Hydropower

Among sources of renewable energy, hydroelectric plants have the advantages of being long-lived—many existing plants have operated for more than 100 years. Also, hydroelectric plants are clean and have few emissions.

3. Solar energy

Solar panels generate electricity by converting photons (packets of light energy) into an electric current.

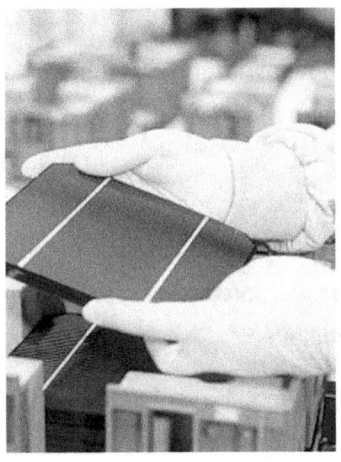

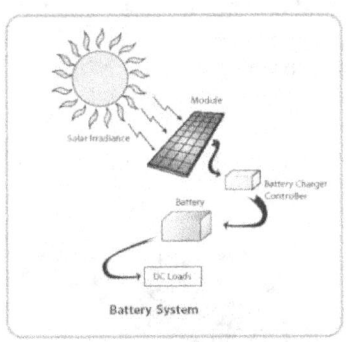

Solar energy is the energy derived from the sun through the form of solar radiation. Solar powered electrical generation relies on photo voltaic and heat engines. A partial list of other solar applications includes space heating and cooling through solar architecture, day lighting, solar hot water, solar cooking, and high temperature process heat for industrial purposes.

Solar technologies are broadly characterized as either passive solar or active solar depending on the way they capture, convert and distribute solar energy. Active solar techniques include the use of photovoltaic panels and solar thermal collectors to harness the energy. Passive solar techniques include orienting a building to the Sun, selecting materials with favorable thermal mass or light dispersing properties, and designing spaces that naturally circulate air.

Strano's nanotube antenna boosts the number of photons that can be captured and transforms the light into energy that can be funneled into a solar cell.

The antenna consists of a fibrous rope about 10 micrometers (millionths of a meter) long and four micrometers thick, containing about 30 million carbon nanotubes.

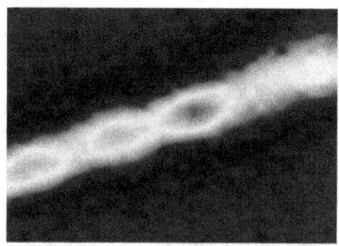

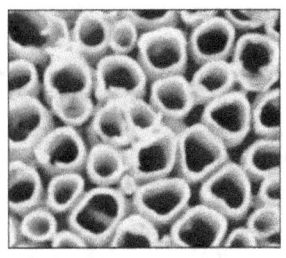

Strano's team built, for the first time, a fiber made of two layers of nanotubes with different electrical properties -- specifically, different band gaps.

In any material, electrons can exist at different energy levels. When a photon strikes the surface, it excites an electron to a higher energy level, which is specific to the material. The interaction between the energized electron and the hole it leaves behind is called an exciton, and the difference in energy levels between the hole and the electron is known as the band gap.

The inner layer of the antenna contains nanotubes with a small band gap, and nanotubes in the outer layer have a higher band gap. That's important because excitons like to flow from high to low energy. In this case, that means the excitons in the outer layer flow to the inner layer, where they can exist in a lower (but still excited) energy state.

When light energy strikes the material, all of the excitons flow to the center of the fiber, where they are concentrated. Strano and his team have not yet built a photovoltaic device using the antenna, but they plan to. In such a device, the antenna would concentrate photons before the photovoltaic cell converts them to an electrical current. This could be done by constructing the antenna around a core of semiconducting material.

The interface between the semiconductor and the nanotubes would separate the electron from the hole, with electrons being collected at one electrode touching the inner semiconductor, and holes collected at an electrode touching the nanotubes. This system would then generate electric current. The efficiency of such a solar cell would depend on the materials used for the electrode, according to the researchers.

Strano's team is the first to construct nanotube fibers in which they can control the properties of different layers, an achievement made possible by recent advances in separating nanotubes with different properties.

While the cost of carbon nanotubes was once prohibitive, it has been coming down in recent years as chemical companies build up their manufacturing capacity. "At some point in the near future, carbon nanotubes will likely be sold for pennies per pound, as polymers are sold," says Strano. "With this cost, the addition to a solar cell might be negligible compared to the fabrication and raw material cost of the cell itself, just as coatings and polymer components are small parts of the cost of a photovoltaic cell."

Strano's team is now working on ways to minimize the energy lost as excitons flow through the fiber, and on ways to generate more than one exciton per photon. The nanotube bundles described in the Nature Materials paper lose about 13 percent of the energy they absorb, but the team is working on new antennas that would lose only 1 percent [2].

4. Biomass

Biomass (plant material) is a renewable energy source because the energy it contains comes from the sun. Through the process of photosynthesis, plants capture the sun's energy. When the plants are burned, they release the sun's energy they contain. In this way, biomass functions as a sort of natural battery for storing solar energy.

As long as biomass is produced sustainably, with only as much used as is grown, the battery will last indefinitely.

In general there are two main approaches to using plants for energy production: growing plants specifically for energy use, and using the residues from plants that are used for other things. The best approaches vary from region to region according to climate, soils and geography

5. Biofuel

Liquid biofuel is usually either bio alcohol such as bioethanol or oil such as biodiesel.

Bioethanol is an alcohol made by fermenting the sugar components of plant material and it is made mostly from sugar and starch crops. With advanced technology being developed, cellulosic biomass, such as trees and grasses, are also used as feed stocks for ethanol production.

Ethanol can be used as a fuel for vehicles in its pure form, but it is usually used as a gasoline additive to increase octane and improve vehicle emissions. Bioethanol is widely used in the USA and in Brazil [3].

6. Geothermal energy

The geothermal energy from the core of the Earth is closer to the surface in some areas than in others. Where hot underground steam or water can be tapped and brought to the surface it may be used to generate electricity.

Such geothermal power sources exist in certain geologically unstable parts of the world such as Chile, Iceland, New Zealand, United States, the Philippines and Italy.

The two most prominent areas for this in the United States are in the Yellowstone basin and in northern California.

Geothermal energy is energy obtained by tapping the heat of the earth itself, both from kilometers deep into the Earth's crust in some places of the globe or from some meters in geothermal heat pump in all the places of the planet. It is expensive to build a power station but operating costs are low resulting in low energy costs for suitable sites. Ultimately, this energy derives from heat in the Earth's core.

Three types of power plants are used to generate power from geothermal energy: dry steam, flash, and binary. Dry steam plants take steam out of fractures in the ground and use it to directly drive a turbine that spins a generator.

Flash plants take hot water, usually at temperatures over 200 °C, out of the ground, and allows it to boil as it rises to the surface then separates the steam phase in steam/water separators and then runs the steam through a turbine.

In binary plants, the hot water flows through heat exchangers, boiling an organic fluid that spins the turbine. The condensed steam and remaining geothermal fluid from all three types of plants are injected back into the hot rock to pick up more heat.

Iceland produced 170 MW geothermal power and heated 86% of all houses in the year 2000 through geothermal energy. Some 8000 MW of capacity is operational in total.

There is also the potential to generate geothermal energy from hot dry rocks. Holes at least 3 km deep are drilled into the earth. Some of these holes pump water into the earth, while other holes pump hot water out.

The heat resource consists of hot underground radiogenic granite rocks, which heat up when there is enough sediment between the rock and the Earth's surface. Several companies in Australia are exploring this technology.

7. Tidal energy

Tidal power can be extracted from Moon-gravity-powered tides by locating a water turbine in a tidal current, or by building impoundment pond dams that admit-or-release water through a turbine.

The turbine can turn an electrical generator, or a gas compressor, that can then store energy until needed. Coastal tides are a source of clean, free, renewable, and sustainable energy.

8. Hydrogen obtained by artificial photosynthesis

Artificial photosynthesis is a research field that attempts to replicate the natural process of photosynthesis, converting sunlight, water, and carbon dioxide into carbohydrates and oxygen.

Sometimes, splitting water into hydrogen and oxygen by using sunlight energy is also referred to as artificial photosynthesis. The actual process that allows half of the overall photosynthetic reaction to take place is photo-oxidation. This half-reaction is essential in separating water molecules because it releases hydrogen and oxygen ions. These ions are needed to reduce carbon dioxide into a fuel. However, the only known way this is possible is through an external catalyst, one that can react quickly as well as constantly absorb the sun's photons. The general basis behind this theory is the creation of an "artificial plant" type fuel source.

Artificial photosynthesis is a renewable, carbon-neutral source of fuel, producing either hydrogen, or carbohydrates. This sets it apart from the other popular renewable energy sources — hydroelectric, solar photovoltaic, geothermal, and wind — which produce electricity directly, with no fuel intermediate.

As such, artificial photosynthesis may become a very important source of fuel for transportation. Unlike biomass energy, it does not require arable land, and so it need not compete with the food supply.

Since the light-independent phase of photosynthesis fixes carbon dioxide from the atmosphere, artificial photosynthesis may provide an economical mechanism for carbon sequestration, reducing the pool of CO_2 in the atmosphere, and thus mitigating its effect on global warming. Specifically, net reduction of CO_2 will occur when artificial photosynthesis is used to produce carbon-based fuel which is stored indefinitely.

9. Blacklight power

Beginning in 1986, Dr. Randell L. Mills developed the theory on which the BlackLight Process is based. In 1989, the original patent applications were filed and the conclusions of the theoretical work were published. Dr. Mills believes that he has succeeded with the unification of gravity with atomic physics. In 1991, Dr. Mills founded Hydro Catalysis Power Corp. to pursue the development and ultimate commercialization of a new form of energy - the Hydro Catalysis Process.

In the fall of 1996, the Company's name was changed from Hydro Catalysis Power Corp. to BlackLight Power, Inc. to reflect the ultraviolet light emission produced by catalysis in the renamed BlackLight Process. In 1999 the Company moved to its present location, a 53,000 square-foot research facility, in Cranbury, NJ, and has since expanded its employee base to 25 people.

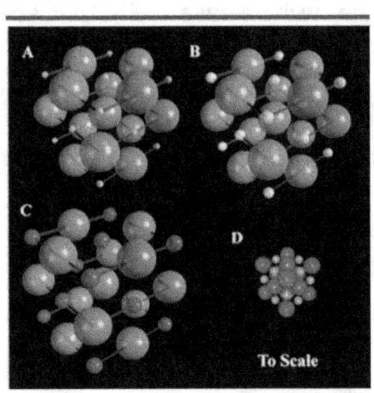

Based on physical laws of nature, Dr. Mills' theory predicts that additional lower energy states are possible for the hydrogen atom, but are not normally achieved.

They are not normally achieved because transitions to these states are not directly associated with the emission of radiation, thus the ordinary hydrogen atom, as well as lower energy hydrogen atoms (termed hydrinos), are stable in isolation.

Mills' theory further predicts that hydrogen atoms can achieve these states by a radiation-less energy transfer with a nearby atom, ion, or combination of ions (a catalyst) having the capability to absorb the energy required to effect the transition. (Radiation-less energy transfer is common. For example, it is the basis of the performance of the most common phosphor used in fluorescent lighting.) Thus, the Company believes hydrogen atoms can be induced to jump to a lower energy state, with release of the net energy difference between states.

Successive stages of collapse of the hydrogen atom are predicted, resulting in the release of energy in amounts many times greater than the energy released by the combustion of hydrogen.

Since the combustion energy is equivalent to the energy required to liberate hydrogen from water, a process, which takes water as a feed material and produces net energy, is possible. The equivalent energy content of water would thus be several hundred to several thousand times that of crude oil, depending on the average number of stages of collapse.

10. Waves power

Wave power is the transport of energy by ocean surface waves, and the capture of that energy to do useful work — for example for electricity generation, water desalination, or the pumping of water (into reservoirs).

Wave power is distinct from the diurnal flux of tidal power and the steady gyre of ocean currents.

Wave power generation is not currently a widely employed commercial technology although there have been attempts at using it since at least 1890.

In 2008, the first experimental wave farm was opened in Portugal, at the Aguçadoura Wave Park.

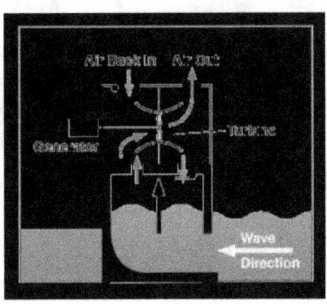

NEW METHODS OF OBTAINING ENERGY

1. Submarines power plants in the future

LONDON: A massive underwater river flowing along the bottom of the Black Sea has been found by scientists - a discovery that could help explain how life manages to survive in the deep oceans away from the nutrient-rich waters found close to land.

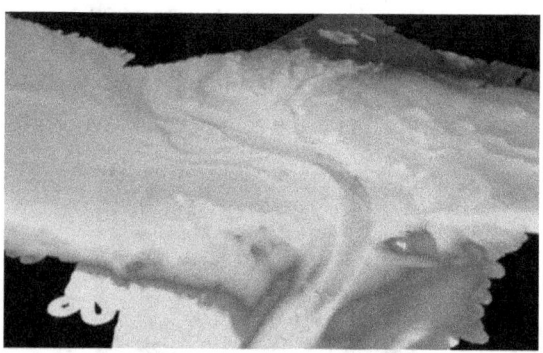

It is estimated that if on land, the undersea river would be the world's sixth largest in terms of the volume of water flowing through it. Researchers working in the Black Sea have found currents of water 350 times greater than the River Thames flowing along the sea bed, carving out channels much like a river on the land, the Telegraph reports.

The undersea river, which is up to 115 feet deep in places, even has rapids and waterfalls much like its terrestrial equivalents.

The scientists, based at the University of Leeds, used a robotic submarine to study a deep channel that had been found on the sea bed, and found a river of highly salty water flowing along the deep channel at the bottom of the Black Sea, creating river banks and flood plains much like a river on land.

Dan Parsons, from the university's School of Earth and Environment, said, "It flows down the sea shelf and out into the abyssal plain much like a river on land.

The abyssal plains of our oceans are like deserts of marine world, but these channels can deliver nutrients and ingredients needed for life out over these deserts. "This means they could be vitally important, like arteries providing life to the deep ocean.

The key difference we found from terrestrial rivers was that as the flow goes round the bend, the water spirals in the opposite way to rivers on land," Parsons said. The undersea river, which is yet to be named, stems from salty water spilling through the Bosphorus Strait from the Mediterranean into the Black Sea, where the water has a lower salt content.

Installation of some turbines that generate electricity in the underwater river which flowing along the bottom of the Black Sea, could bring to Europe a large amount of cheap and clean energy.

2. Obtaining energy with alpha Stirling engines

We can try the Alpha Stirling Motors for to obtain energy from two locations with different temperatures, ground and underground for example.

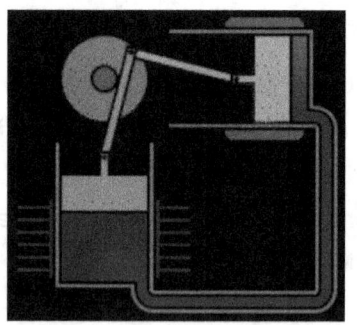

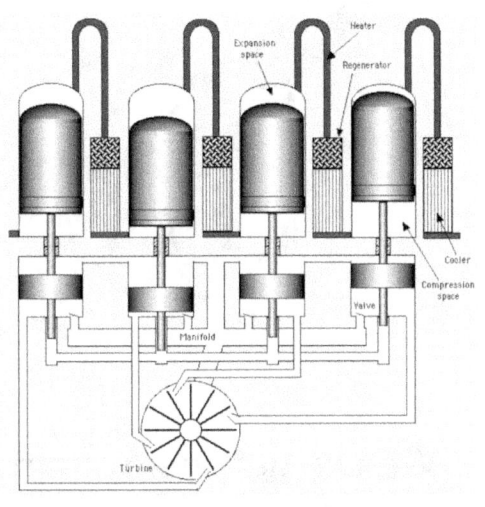

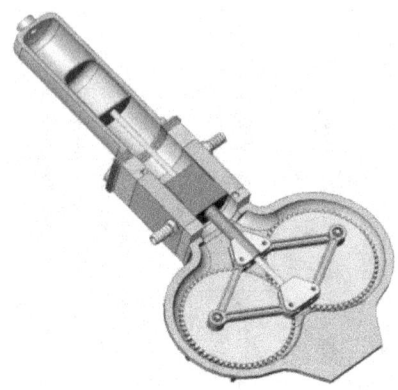

3. We can get energy from inside a volcano

We will install various pipes, serpentines, boilers, inside of some volcanoes, and by pumping the cold water in them we will obtain hot water to the outer.

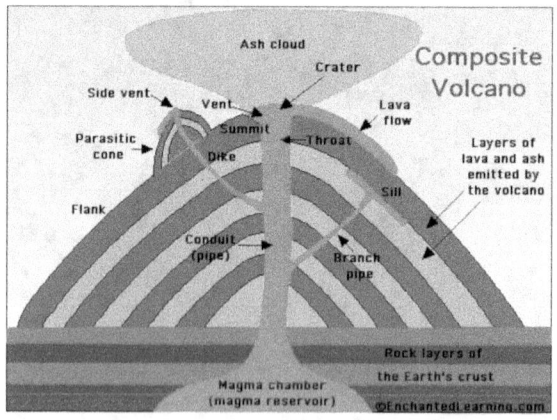

4. Capture and keeping of the energy liberated by a lightning

The lightning has an power of 3000000000000[W] =$3*10^{12}$[W] =$3*10^{9}$[kW] =$3*10^{6}$[MW] =$3*10^{3}$[GW].

Lightning is produced at Earth surface with an average of 300 kicks per second. If we could collect and keep all this energies, who are liberated by a single lightning, we could obtain 1-7 GJ=1-7 GWs/second, 1-7 GWh/hour=8760-61320 GWh/year.

The lightning can be attracted and retained by huge balls buried in the planet's surface, in places where are more frequent rains.

The areas chosen should be as well insulated and removed, to prevent unauthorized access inside them.

5. Extracting energy from electron

Getting energy, renewable, clean, friendly (not dangerous), cheaper, by annihilation (For example, the annihilation of an electron with an antielectron).

Electron and positron are obtained by extracting them from atoms; the extraction, consume a negligible amount of energy.

Then, the two particles are brought near one another (collision); now it occur the phenomenon of annihilation, when the rest mass is converted totally into energy (gamma photons).

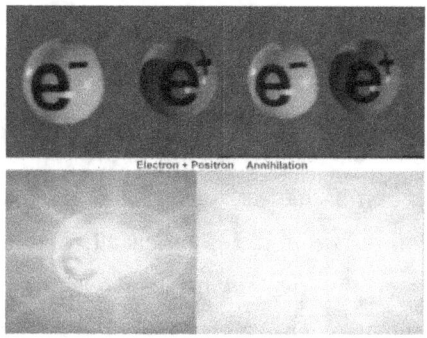

Occur gamma photons, as many as needed to retrieve the total energy of the electron and positron (rest energy and kinetic energy); usually one can get two or three gamma particles (when we have a lower annihilation, ie two antiparticles with lower energy, each with a little beyond rest mass, ie the particles are accelerated at a low-speed motion), but we can get more particles when we have a high annihilation (ie when the particle energy is high and the particles were strongly accelerated before the collision).

Rest energy of an electron-positron pairs exceeds slightly 1 MeV (what is an extremely large energy from some as small particles, comparable energy with that achieved by the merger of two much larger particles, having rest mass of about 2000 times higher).

Hence the first great advantage of the new method proposed, namely that if the most complex physical phenomenon so far tried to get inside the material energy (hot or cold fusion), draw only about a thousandth part of the rest mass of the particle, resulting in the fusion of two particles practically only the energy gap between energy particles being free and their energy when they are united, the proposed method to extract virtually all the internal energy of the particles annihilated.

We started with the electron positron pair because these small particles are more easily extracted from the atoms (the atoms are then immediately regenerated naturally, which determines the nature of renewable energy from the annihilation of particles).

Next step is to test the annihilation between a proton and an antiproton, because their mass is about 1800 times higher than that of the electron and positron, resulting in their annihilation as an energy by about 1000 times higher, ie instead of 1 MeV, 1 GeV (is considered as the only real obtained energy, the energy donated by the proton of the hydrogen ion; but the energy of an antiproton is considered to be donated by us almost entirely, for now, because to obtain today an antiproton we must accelerate some particles at very high-energy and then collide them).

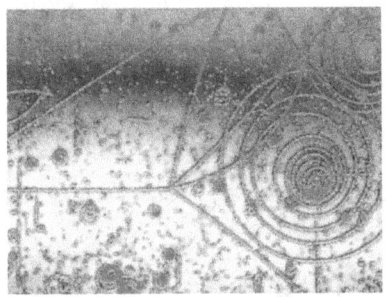

So the real comparison must to be made between the deuterons fusion and annihilation process of a hydrogen ion (proton) with an antiproton. It will be a difference of energy of about 1000 times higher per pair of particles used, in favor of the annihilation process.

Practically it realizes the dream of extracting energy from all the matter.

Another great advantage of this method is that no radioactive substances and are not radioactive wastes from the process. From this process we obtain only gamma photons (ie energy) and possibly other energetic mini particles.

The process does not pose any threat to humans and the environment.

The energy produced is clean. The technology required is much simpler than nuclear (fission or fusion), cheaper and easier to maintain. Enough energy is given by the annihilation process (virtually unlimited), cheap, clean, safe, renewable immediately (sustainable), with technology made simple.

We can extract the energy of the rest mass of an electron. For a pair of an electron and a positron this energy is circa 1 MeV.

The "synchrotron radiation (synchrotron light source)" produces deliberated a radiation source.

Electrons are accelerated to high speeds in several stages to achieve a final energy (that is typically in the GeV range).

We need two synchrotrons, a synchrotron for electrons and another who accelerates positrons.

The particles must to be collided, after they are being accelerated to an optimal energy level.

All the energies are collected at the exit of the Synchrotrons, after the collision of the opposite particles.

We will recover the accelerating energy, and in addition we also collect the rest energy of the electrons and positrons.

At a rate of 10^{19} electrons/s we obtain an energy of about 7 GWh / year, if even are produced only half of the possible collisions.

This high rate can be obtained with 60 pulses per minute and 10^{19} electrons per pulse, or with 600 pulses per minute and 10^{18} electrons per pulse.

If we increase the flow rate of 1,000 times, we can have a power of about 7 TWh / year. This type of energy can be a complement of the fusion energy, and together they must replace the energy obtained by burning hydrocarbons. Advantages of the annihilation of an electron with a positron, compared with the nuclear fission reactors, are disposal of radioactive waste, of the risk of explosion and of the chain reaction.

Energy from the rest mass of the electron is more easily controlled compared with the fusion reaction, cold or hot.

Now, we don't need of enriched radioactive fuel (as in nuclear fission case), by deuterium, lithium and of accelerated neutrons (like in the cold fusion), of huge temperatures and pressures (as in the hot fusion), etc.

Results and Discussion

How much energy, can we get from inside of the matter? Einstein has showed that from one kg of matter we could get the energy needs for entire Earth for a year:

$$E = m \cdot c^2 = 1[kg] \cdot (3 \cdot 10^8)^2 [(m/s)^2] = 9 \cdot 10^{16} [j] = 2,5 \cdot 10^{10} [KWh] = 2,5 \cdot 10^7 [MWh] = 2,5 \cdot 10^4 [GWh] = 25 [TWh]$$

We could do this, but only if we could extract all the energy from inside the matter.

Through nuclear fusion reaction can be extracted only a part of the rest energy of the particles used. This drop of energy (1 / 1000 of the mass energy of a proton-neutron pairs) is called, discrepancy.

For a kg of particles proton-neutron pairs, fusion energy is about a thousand times smaller than the total energy of a kilogram of matter (only 29 [GWh] from the total internal energy, 25 [TWh]); and considering that a return of 100% fusion reaction, which can't be done anyway.

Theoretically speaking, we can't draw from within the matter (through nuclear fusion reaction) than at most the thousandth part of its energy. Having in view the yield of the nuclear fusion reaction, this obtained energy is and less.

Through reaction of nuclear fission, the energies obtained will be even smaller.

The solution proposed in this work, obtaining energy by the mutual annihilation of two opposite particles, makes possible the requirement of extracting whole energy contained in matter.

A pair formed by a particle and its antiparticle, are brought side by side, at a distance which allow the process of reciprocal annihilation.

To increase the yield of the annihilation reaction (the number of annihilated particles from all particles that exist), we can accelerate the particles and antiparticles separately, and then we may send them into a room where they encounter annihilation at speeds and energies high, or at velocities and energies very high.

If we use electrons and positrons for the reaction of annihilation, it results photons of the gamma type.

In this case, to prevent the possible decay of the obtained photons, again into electrons and positrons (for beginning of this annihilation process with success), the antiparticles and particles used in the process of annihilation, should be collided at low speeds and with low energy.

We can test then the optimum energy particle which permits the reaction with the maxim yield. It is necessary that most particles and antiparticles used, to meet and annihilate each other, and it should be stable as many of the obtained gamma particles.

Conclusions

The fission energy was a necessary evil. In this mode it stretched the oil life, avoiding an energy crisis.

Even so, the energy obtained from hydrocarbons represents today about 66% of all energy used.

At this rate of use of oil, it will be consumed in about 40 years. Today, the production of energy obtained by nuclear fusion is not yet perfect prepared. But time passes quickly. We must rush to implement of the additional sources of energy already known, but and find new energy sources. In these conditions the proposed method to obtaining energy by annihilation of matter and antimatter, can be a real alternative sources of renewable energy.

6. Ocean wave energy captured directly from surface waves

Ocean wave energy can be captured directly from surface waves.

Blowing wind and pressure fluctuations below the surface are the main reasons for causing waves.

But consistency of waves differs from one area of ocean to another. Some regions of oceans receive waves with enough uniformity and force.

Ocean waves contain tremendous energy. Currently scientists and companies are considering exploiting the wave power of oceans to harness clean and green energy.

ANSYS Inc is a global trend setter of simulation software and technologies. Recently it has developed software that is assisting in converting the persistent forces of ocean waves into electricity. Green Ocean Energy Ltd is an Aberdeen based renewable energy company. Their mission is to create and innovate in the field of clean and green energy. They are developing mechanisms to harness energy from the Earth's oceans. They are also focusing on other facts such as the economic viability and sustainability of their products.

Green Ocean Energy has produced two innovative devices – Ocean treader and Wave treader. These devices will move on the ocean surface in a manner as if someone is nodding, these up and down movements of the arms will help in generating power.

It is estimated that each machine will produce around 500KW of electrical power. This power can be transported to the shore with the aid of underwater cables. This amount of electricity can be utilized by about 125 homes. Wave power farms can be established to generate any required amount of power.

Ocean Treader is a floating device. It will be tied up 1 – 2 miles offshore in ocean wave systems. It will not pose any obstruction on the shoreline. The theory has been put to test in wave tank. Now the company is producing a full size machine for offshore testing.

Wave Treader has grown out of our work with Ocean Treader. Wave Treader uses its Sponsons and Arms and are mounted on the base of a static offshore structure. That structure can be a Wind Turbine or Tidal Turbine. By sharing the high infrastructure costs with another device, such as the foundation costs, cabling costs, etc., the economics of both devices are enhanced and the energy yield for a given sea area greatly improved.

Green Ocean Energy has receives a noteworthy boost to the development of its wave power technology after managing £100,000 of funding from the Scottish Enterprise Seed Fund. They have also secured £150,000 of private investment. Graeme Bell, Special Projects Director at Green Ocean Energy said: "We are delighted to receive this support from Scottish Enterprise. The funding will enable the company to take a major step forward and begin detailed engineering and design of a full scale Wave Treader. It's been an exciting time for the company and we're enjoying a fantastic level of interest in our activities."

This financial support will enable the company to continue the engineering and testing of its ground breaking Wave Treader device. This device is affixed to the transition piece of an offshore wind turbine thereby providing combined wind and wave energy.

It is expected that manufacture of a full scale prototype will start next year once an appropriate site has been acknowledged with deployment in early 2011. Commercialization is expected to being in 2012.

(from 2 New & Innovative Ocean Wave Energy Devices-October 5th, 2009, Alternative Energy, in http://www.energyplanet.info/)

7. Scientists at US Laboratory Ready to Create Fusion Energy

The long quest for fusion energy is about to pass a milestone. Scientists at the Lawrence Livermore National Laboratory in California say they are ready to test a giant laser system.

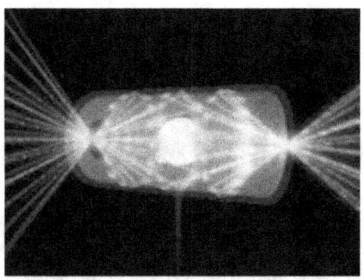

The laser beams expect to start a nuclear reaction that could lead to practical production of energy with an infinite fuel source and no carbon waste.

The goal is to produce energy the same way the sun does - efficiently, cleanly and infinitely.

It is called fusion - and that is what these scientists at the Lawrence Livermore National Laboratory in California are trying to achieve.

Ed Moses directs the Lab's National Ignition Facility, or NIF. "Fusion energy is the long term solution," Moses explains. "It is infinite, essentially infinite fuel. And it has no carbon waste."

"Sun in a Bottle" is how science writer Charles Seife describes fusion.

"Basically all life comes eventually from fusion, the fusion of the sun," he said. "And so if we could replicate this on earth, we've got the clean elemental power that powers everything on earth essentially."

Using a newly-completed $4 billion laser system that took 10 years to build, scientists at NIF say they are very close to producing controlled fusion. If successful, the experiment could be a giant step towards satisfying the world's increasing demands for energy. They say they expect to start experiments next month that they believe will lead to "ignition."

"When we talk about ignition, we talk about making a small sun on earth; getting thermonuclear burn in the laboratory in a controlled manner," Moses clarified.

That burn, or fusion, is what happens in the sun, stars, and in hydrogen bombs: atoms are merged together and fused at very high temperatures, producing enormous amounts of energy.

Scientists at the National Ignition Facility will focus the intense energy of 192 giant laser beams on a hydrogen-filled target the size of a pill, in order to fuse, or ignite, the nuclei of the hydrogen atoms.

"It will make energy as Einstein told us; the neutron will fly off and if we collect the energy that's coming out of that, we have fusion energy to use for electricity or a variety of other purposes," Moses said.

But Charles Seife cautions it will be important for the project to generate more energy than it uses. "But is this going to actually produce more energy than the laser consumes?" he questioned.

"Almost certainly not. And in fact, if you press them, they hedge a little bit, maybe we'll get ignition, maybe we won't."

For now, the focus is on ignition. In the future, the scientists at NIF hope fusion will be a long term energy solution, with the goal of creating more power than they use.

"It is the silver bullet, Moses said. "Of course, there's a lot to do along the way."

Moses says he thinks pilot fusion power plants will be running within a dozen years, and he adds, soon after that, fusion power will be part of our lives.

(From "Voice of America", San Francisco, CA   February 05, 2009)

Nuclear fusion plant possible within a decade, physicist says.

Using the most powerful laser system ever built, scientists have brought us one step closer to nuclear fusion power, a new study says.

The same process that powers our sun and other stars, nuclear fusion has the potential to be an efficient, carbon-free energy source—with none of the radioactive waste associated with the nuclear fission method used in current nuclear plants.

(See "Radioactive Rabbit Droppings Help Spur Nuclear Cleanup.")

Thanks to the new achievement, a prototype nuclear fusion power plant could be operating within a decade, speculated study leader Siegfried Glenzer, a physicist at Lawrence Livermore National Laboratory in California.

Glenzer and colleagues used the world's largest laser array—the Livermore lab's National Ignition Facility—to heat a BB-size fuel pellet to millions of degrees Fahrenheit.

"These lasers are pulsed, and for a very short amount of time"—one ten-billionth of a second—"the power they produce is more than all the power generated by the entire electrical grid of the United States" at any given moment, Glenzer said.

The test confirmed that a technique called inertial fusion ignition could be used to trigger nuclear fusion—the merging of the nuclei of two atoms of, say, hydrogen—which can result in a tremendous amount of excess energy. Nuclear fission, by contrast, involves the splitting of atoms.

The laser demonstration means scientists are now much closer to triggering nuclear fusion in a controlled setting—something that's never been done before and which is necessary if fusion is to be harnessed for energy.

Nuclear's Nice Side?

Performing nuclear fusion in the lab requires enormous amounts of laser power, but if perfected, controlled fusion should generate ten to a hundred times more electrical energy than is used to spark the nuclear reactions. Nuclear fusion, after all, is what allows stars to burn for billions of years.

And fusion could be not only powerful but clean and green as well.

Not only does nuclear fusion not produce long-lasting nuclear waste, but fusion could potentially be used to chemically neutralize radioactive pollutants and has been "proposed as a cure to our nuclear waste problem," Glenzer said.

Simply put, neutrons released by fusion could rearrange radioactive atoms so they aren't radioactive anymore.

(Related: "'Nuclear Archaeologists' Find World War II Plutonium.")

Nuclear fusion energy is also potentially carbon free, meaning it could be used to generate power without creating any more carbon dioxide gas, which contributes to global warming.

And while fossil fuels, such as oil and coal, and nuclear fission fuels, such as uranium, are limited resources, there's enough nuclear fusion fuel on, in, and around our planet "to power the Earth longer than the lifetime of the sun," Glenzer said.

(Related: "Cheap Oil to Last, 'Doomsday' Fears Overblown, Author Says.")

Gold Fusion

During the laser experiment, the fuel pellet was placed inside a solid-gold cylinder about the size of a pencil eraser, which was hit by multiple laser beams.

The gold cylinder absorbed the laser energy and converted it into thermal x-ray energy.

The x-rays then ricocheted inside the cylinder and struck the fuel pellet from all sides. As the pellet absorbed the x-rays, it heated up—eventually reaching about 60 million degrees Fahrenheit (33 million degrees Celsius)—then collapsed in on itself.

The experiment was designed only to test the lasers' ability to heat the cylinder efficiently. Made largely of plastics and helium, the fuel pellet was not filled with enough actual fuel—chemical variants of hydrogen called deuterium and tritium—to actually trigger nuclear fusion.

Actual fusion, Glenzer said, will occur sometime this year.

With a fully loaded fuel pellet, "the implosion will be like squeezing a soccer ball to the size of a pinhead," he added. "The center of that spherical ball will get so hot that nuclear fusion starts."

Nuclear Fusion Plant by 2020?

If successful, the upcoming nuclear fusion experiment will create two classes of energetic particles: alpha particles and neutrons.

"The neutrons escape and can be used to do things like heat up water"—which could potentially be used to produce steam to drive turbines in an electrical plant, Glenzer said.

"The alpha particles remain trapped [in the burning sphere] and continue to heat the fuel and make it burn," as happens in a star.

Scientists estimate that if they can get to the point where they can burn about five fuel pellets a second, a power plant could continuously generate up to a gigawatt of energy—about what the city of San Francisco is consuming at any given moment.

A working prototype of a such a plant could be built in a decade, Glenzer said.

Cheaper to Burn Cash?

Nuclear fusion researcher Michael Mauel is "very excited" about the recent experiment and said it shows the ignition method works as expected.

But "whether or not we'll have lasers imploding pellets to make fusion energy—it's way too early to tell," said Mauel, who was not involved in the study, which will be published in the journal Science tomorrow.

In addition to the considerable engineering challenges involved in ramping up the laser systems for wide-scale use, the cost of the fuel pellets will also have to come down, said Mauel, a Columbia University physicist.

"Each one of these costs between ten [thousand] and a hundred thousand dollars," Mauel said. To use the pellet method to generate nuclear fusion power, "they'll have to cost less than ten cents a piece."

(From http://news.nationalgeographic.com/news/2010/01/100128-nuclear-fusion-power-lasers-science/)

The origin of the energy released in fusion of light elements is due to an interplay of two opposing forces, the nuclear force which draws together protons and neutrons, and the Coulomb force which causes protons to repel each other.

The protons are positively charged and repel each other but they nonetheless stick together, portraying the existence of another force referred to as a nuclear attraction. The strong nuclear force, that overcomes electric repulsion in a very close range.

The effect of this force is not observed outside the nucleus. Hence the force has a strong dependence on distance making it a short range force.

The same force also pulls the neutrons together, or neutrons and protons together. Because the nuclear force is stronger than the Coulomb force for atomic nuclei smaller than iron and nickel, building up these nuclei from lighter nuclei by fusion releases the extra energy from the net attraction of these particles.

For larger nuclei, however, no energy is released, since the nuclear force is short-range and cannot continue to act across still larger atomic nuclei. Thus, energy is no longer released when such nuclei are made by fusion (instead, energy is absorbed in such processes).

Fusion reactions of light elements power the stars and produce virtually all elements in a process called nucleosynthesis.

The fusion of lighter elements in stars releases energy (and the mass that always accompanies it).

For example, in the fusion of two hydrogen nuclei to form helium, seven-tenths of 1 percent of the mass is carried away from the system in the form of kinetic energy or other forms of energy (such as electromagnetic radiation). However, the production of elements heavier than iron absorbs energy.

Research into controlled fusion, with the aim of producing fusion power for the production of electricity, has been conducted for over 60 years. It has been accompanied by extreme scientific and technological difficulties, but has resulted in progress.

At present, controlled fusion reactions have been unable to produce break-even (self-sustaining) controlled fusion reactions.

Workable designs for a reactor that theoretically will deliver ten times more fusion energy than the amount needed to heat up plasma to required temperatures (see ITER) were originally scheduled to be operational in 2018, however this has been delayed and a new date has not been stated.

It takes considerable energy to force nuclei to fuse, even those of the lightest element, hydrogen.

This is because all nuclei have a positive charge (due to their protons), and as like charges repel, nuclei strongly resist being put too close together.

Accelerated to high speeds (that is, heated to thermonuclear temperatures), they can overcome this electrostatic repulsion and get close enough for the attractive nuclear force to be sufficiently strong to achieve fusion.

The fusion of lighter nuclei, which creates a heavier nucleus and often a free neutron or proton, generally releases more energy than it takes to force the nuclei together; this is an exothermic process that can produce self-sustaining reactions.

The US National Ignition Facility, which uses laser-driven inertial confinement fusion, is thought to be capable of break-even fusion.

Energy released in most nuclear reactions is much larger than in chemical reactions, because the binding energy that holds a nucleus together is far greater than the energy that holds electrons to a nucleus.

For example, the ionization energy gained by adding an electron to a hydrogen nucleus is 13.6 eV—less than one-millionth of the 17 MeV released in the deuterium–tritium (D–T) reaction shown in the diagram to the right.

Fusion reactions have an energy density many times greater than nuclear fission; the reactions produce far greater energies per unit of mass even though individual fission reactions are generally much more energetic than individual fusion ones, which are themselves millions of times more energetic than chemical reactions.

Only direct conversion of mass into energy, such as that caused by the annihilation collision of matter and antimatter, is more energetic per unit of mass than nuclear fusion.

A substantial energy barrier of electrostatic forces must be overcome before fusion can occur. At large distances two naked nuclei repel one another because of the repulsive electrostatic force between their positively charged protons.

If two nuclei can be brought close enough together, however, the electrostatic repulsion can be overcome by the attractive nuclear force, which is stronger at close distances.

When a nucleon such as a proton or neutron is added to a nucleus, the nuclear force attracts it to other nucleons, but primarily to its immediate neighbours due to the short range of the force.

The nucleons in the interior of a nucleus have more neighboring nucleons than those on the surface.

Since smaller nuclei have a larger surface area-to-volume ratio, the binding energy per nucleon due to the nuclear force generally increases with the size of the nucleus but approaches a limiting value corresponding to that of a nucleus with a diameter of about four nucleons.

It is important to keep in mind that the above picture is a toy model because nucleons are quantum objects, and so, for example, since two neutrons in a nucleus are identical to each other, distinguishing one from the other, such as which one is in the interior and which is on the surface, is in fact meaningless, and the inclusion of quantum mechanics is necessary for proper calculations.

The electrostatic force, on the other hand, is an inverse-square force, so a proton added to a nucleus will feel an electrostatic repulsion from all the other protons in the nucleus.

The electrostatic energy per nucleon due to the electrostatic force thus increases without limit as nuclei get larger.

H-hour

With the help of powerful lasers one can create a dense and highly ionized plasma. We need a highly ionized dense plasma to achieve nuclear fusion (cold or hot).

Since 1989, it talks about achieving nuclear fusion hot and cold. Another two decades have passed and humanity still does not benefit from nuclear fusion energy.

What actually happens? Is it an unattainable myth? It was also circulated by the media that has been achieved nuclear fusion heat. Since 1989 there are all sorts of scientists with all kinds of crafted devices, which declare that they can produce nuclear power obtained by cold fusion (using cold plasma).

May be that these devices works, but their yield is probably too small, or at an enlarged scale these give not the expected results. This is the real reason why we can't use yet the survival fuel (the deuterium).

Unfortunately today the dominant processes that produce energy are combustion (reaction) chemical combination of carbon with oxygen. Thermal energy released from such reactions is conventionally valued at about 7000 calories per gram.

Only the early 20th century physicists have succeeded in producing, other energy than by traditional methods. Energy release per unit mass was enormous compared with that obtained by conventional procedures.

The Kilowatt based on nuclear fission of uranium nuclei has today a significant share in global energy balance.

Unfortunately, the nuclear power plants burn the fuel uranium, already considered conventional and on extinct.

The current nuclear power is considered a transition way, to the energy thermonuclear, based on fusion of light nuclei.

The main particularity of synthesis reaction (fusion) is the high prevalence of the used fuel (primary), deuterium. It can be obtained relatively simply from ordinary water.

Deuterium was extracted from water for the first time by Harold Urey in 1931. Even at that time, small linear electrostatic accelerators, have indicated that D-D reaction (fusion of two deuterium nuclei) is exothermic.

Today we know that not only the first isotope of hydrogen (deuterium) produces fusion energy, but and the second (heavy) isotope of hydrogen (tritium) can produce energy by nuclear fusion.

The first reaction is possible between two nuclei of deuterium, from which can be obtained, either a tritium nucleus plus a proton and energy, or an isotope of helium with a neutron and energy.

$$_1^2D + {_1^2D} \rightarrow \begin{cases} _1^3T + 1MeV + {_1^1H} + 3MeV = {_1^3T} + {_1^1H} + 4MeV & (1) \\ _2^3He + 0.8MeV + {^1n} + 2.5MeV = {_2^3He} + {^1n} + 3.3MeV & (2) \end{cases}$$

Observations: a deuterium nucleus has a proton and a neutron; a tritium nucleus has a proton and two neutrons.

Fusion can occur between a nucleus of deuterium and one of tritium.

$${}_1^2D + {}_1^3T \to {}_2^4He + 3.5 MeV + {}^1n + 14 MeV = {}_2^4He + {}^1n + 17.5 MeV \quad (3)$$

Another fusion reaction can be produced between a nucleus of deuterium and an isotope of helium.

$${}_1^2D + {}_2^3He \to {}_2^4He + 3.7 MeV + {}_1^1H + 14.7 MeV = {}_2^4He + {}_1^1H + 18.4 MeV \quad (4)$$

For these reactions to occur, should that the deuterium nuclei have enough kinetic energy to overcome the electrostatic forces of rejection due to the positive tasks of protons in the nuclei.

For deuterium, for average kinetic energy are required tens of keV.

For 1 keV are needed about 10 million degrees temperature. For this reason hot fusion requires a temperature of hundreds of millions of degrees.

The huge temperature is done with high power lasers acting hot plasma.

Electromagnetic fields are arranged so that it can maintain hot plasma.

The best results were obtained with the Tokamak-type installations.

ITER: the world's largest Tokamak

ITER is based on the 'tokamak' concept of magnetic confinement, in which the plasma is contained in a doughnut-shaped vacuum vessel. The fuel—a mixture of deuterium and tritium, two isotopes of hydrogen—is heated to temperatures in excess of 150 million°C, forming a hot plasma. Strong magnetic fields are used to keep the plasma away from the walls; these are produced by superconducting coils surrounding the vessel, and by an electrical current driven through the plasma.

Deuterium fuel is delivered in heavy water, D_2O.
Tritium is obtained in the laboratory by the following reaction.

$$_3^6Li + {}^1n \rightarrow {}_1^3T + {}_2^4He + 4.6 MeV \quad (5)$$

Lithium, the third element in Mendeleev's table, is found in nature in sufficient quantities.
The accelerated neutrons which produce the last presented reaction with lithium, appear from the second and the third presented reaction.
Raw materials for fusion are deuterium and lithium.

All fusion reactions shown produce finally energy and He. He is a (gas) inert element. Because of this, fusion reaction is clean, and far superior to nuclear fission.

Hot fusion works with very high temperatures.

In cold fusion, it must accelerate the deuterium nucleus, in linear or circular accelerators.

Final energy of accelerated deuterium nuclei should be well calibrated for a positive final yield of fusion reactions (more mergers, than fission).

Electromagnetic fields which maintain the plasma (cold and especially the warm), should be and constrictors (especially at cold fusion), for to press, and more close together the nuclei.

The potential energy with that two particles reject each other, can be approximately calculated with the following relationship.

$$U \equiv E_p = \frac{1}{4\pi\varepsilon_0} \cdot \frac{q_1 \cdot q_2}{d_{12}} = \frac{1}{4\pi \cdot 8.8541853 \cdot 10^{-12}} \cdot \frac{(1.602 \cdot 10^{-19})^2}{4 \cdot 10^{-15}} =$$
$$= 5.7664 \cdot 10^{-14} [J] = 5.7664 \cdot 10^{-14} \cdot 6.242 \cdot 10^{18} [eV] = 3.599 \cdot 10^5 [eV] =$$
$$= 360 [keV] \qquad (6)$$

At a keV is necessary a temperature of 10 million ^{0}C.

At 360 keV is necessary a temperature of 3600 million ^{0}C.

In hot fusion it need a temperature of 3600 million degrees.
Without a minimum of 3000 million degrees we can't make the hot fusion reaction, to obtain the nuclear power.

Today we have just 150 million degrees made.

To replace the lack of necessary temperature, it uses various tricks.

In cold fusion one must accelerate the deuterium nuclei at an energy of 360 [keV], and then collide them with the cold fusion fuel (heavy water and lithium).

Cold Nuclear Fusion

Because obtaining the necessary huge temperature for hot fusion is still difficult, it is time to focus us on cold nuclear fusion.
We need to bomb the fuel with accelerated deuterium nuclei.
The fuel will be made from heavy water and lithium.
The optimal proportion of lithium will be tested.
It would be preferable to keep fuel in the plasma state.
Between deuterium and tritium the smallest radius is the radius of deuterium nucleus.

$Deuterium \quad A=2 \quad A^{1/3}=1.259921 \Rightarrow R_D=1.8268855223476 \cdot 10^{-15}[m]$

$Tritium \quad A=3 \quad A^{1/3}=1.44224957 \Rightarrow R_T=2.0912618769457 \cdot 10^{-15}[m]$

We calculate the minimum distance between two particles which meet together.
This is just the diameter of a deuterium nucleus, d_{12D}.

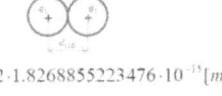

$$d_{12D} = 2 \cdot R_D = 2 \cdot 1.8268855223476 \cdot 10^{-15}[m] =$$
$$= 3.6537710446952 \cdot 10^{-15}[m] =$$
$$\approx 3.653771 \cdot 10^{-15}[m]$$

The deuterium nuclei which will bomb the nuclear fuel, will be accelerated with the (least) energy which reject the two neighboring deuterium nuclei (see the below relationship).

$$U \equiv E_p = \frac{1}{4 \cdot \pi \cdot \varepsilon_0} \cdot \frac{q_1 \cdot q_2}{d_{12D}} = \frac{1}{4 \cdot \pi \cdot 8.8541853 \cdot 10^{-12}} \cdot \frac{(1.602 \cdot 10^{-19})^2}{3.653771 \cdot 10^{-15}} =$$

$$= 6.3128464855 \cdot 10^{-14}[J] = 6.3128464855 \cdot 10^{-14} \cdot 6.242 \cdot 10^{18}[eV] =$$

$$= 3.94 \cdot 10^5[eV] = 3.94 \cdot 10^2[keV] = 394[keV]$$

8. Solar Greenhouse to Produce Food and Electricity

Imagine a greenhouse that is producing solar power and food too. This excellent experiment is being done in Italy. The companies responsible for this project are Renewable energy company Solar ReFeel, CeRSAA and solar panel manufacturer Solyndra. The test site has been constructed at CeRSAA's Albenga, Italy.

The project intends to attain the production of both food and electricity. The research team also wants to validate the crop growth benefits of Solyndra's technology by taking help of independent testing by a leading agricultural research institution.

(From http://www.alternative-energy-news.info/solar-greenhouse-food-electricity/)

Making Ground Source Geothermal a Win-Win Resource for Utilities and Customers

In its 75-year history, modern ground source geothermal energy (GSGE) has flown so far under the radar, it might as well lie in your granddad's root cellar. But unlike root cellars, built as crude geothermal systems to preserve perishables in a static environment, a ground source geothermal heat pump (GSGHP) can deliver a dynamic and effective heating or cooling system.

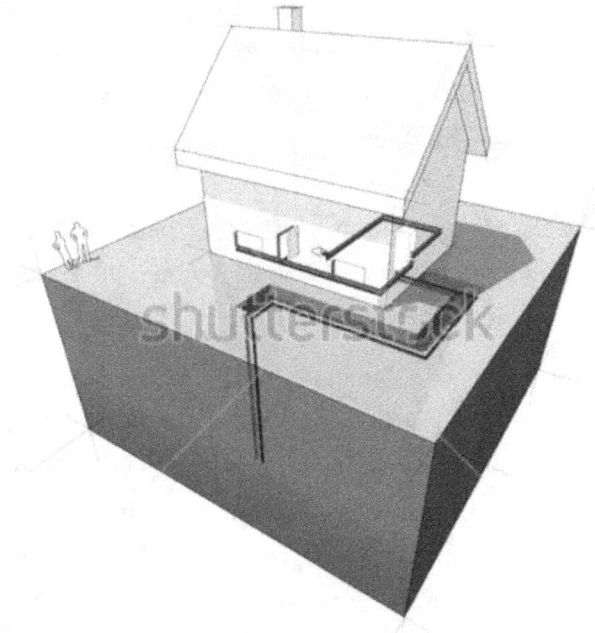

In winter, GSGE extracts heat from the constant temperature reservoir provided by the earth's underground geothermal gradient. Then in summer, the cycle is reversed and GSGHPs reject heat back into the ground.

GSGHPs currently make up some 5 percent of the total heating, ventilation and air conditioning (HVAC) market. In the U.S., an estimated 100,000 residential and commercial GSGHP units are installed annually, with the current market fairly evenly distributed between residential and commercial sectors.

"Ground source geothermal energy is one of the best renewables, even though it took a while for people to consider it a renewable energy source," said Garen Ewbank, an industrial engineer and the owner of Ewbank Geo Testing in Fairview, Oklahoma. "It's renewable because of the constant temperature due to the earth's geothermal gradient. About 1 percent actually comes from solar; the rest is geothermal."

The key to it all is sheer geothermal underground dynamics. A couple of dozen feet under, temperatures remain the same almost year-round. As Ewbank notes, ground source temperatures tend to vary from 49 degrees in Toronto to 72 degrees in the Florida Keys, which he says represents a surprisingly narrow range for such a wide geographic area.

GSGE taps into this constant energy gradient in both winter and summer to heat and cool homes, businesses, schools, hospitals, retail stores, hotels, ski resorts, even the Statue of Liberty gift shop. But the residential market arguably could offer the GSGE industry the most potential growth.

A typical closed-loop installation involves three vertical 250 ft. deep bore holes which are, in turn, loaded with high density polyethylene pipe to circulate water from the indoor heat pump throughout a continuous pressurized loop.

Using a fan, compressor and indoor heat pump, the GSGHP system extracts heat from this constant ground source and during winter compresses it to higher temperatures. During summer, the system simply reverses. The most commonly used GSGHP setup uses a closed-loop heat exchanger, water and underground piping to either extract or reject the heat.

As Ewbank explains, the whole point of the ground source heat pump is to transfer heat against the direction it would naturally go.

Even so, electricity, water and refrigerant are still needed to run the system, which is typically four times as efficient as a conventional HVAC.

The first GSGHP systems in the U.S. were installed in the late 1930s, but fizzled out in the early 1950s due to construction and design issues. But the industry got a reboot in the late 1970s; about a decade after air source heat pumps became common in the U.S.

"Instead of exchanging energy from the environment where you have a refrigerant to air heat exchanger, GSGHPs have a refrigerant to water heat exchanger," said Jeff Spitler, a mechanical engineer at Oklahoma State University in Stillwater.

A fan and duct system then circulates the hot or cold air from the heat exchanger to points throughout the home or business.

But to install and equip a three bedroom house with a GSGHP, count on costs that are roughly double that of a conventional heating and cooling system. However, with tax credits for installation, homeowners can expect to recoup their investment within five years.

"The market is still maturing because there are still places where you can't find a competitive bid on this kind of technology," said Spitler. "And as you go farther south, it's less likely to be economically feasible."

Spitler notes ideal GSGE locations are regions with big temperature swings like Oklahoma.

Ewbank says that in Oklahoma, with brutal 115 degree summers and 0 degree winters, GSGHPs can tap into a constant deep earth temperature of about 64 degrees. The end result is that ground source energy can be pumped in and out of houses or businesses, using less refrigerant and less power than conventional HVAC systems.

"Ground Source Geothermal is a more reliable and efficient way of doing air conditioning," said Eric Woodroof, the Kentucky-based CEO of Profitable Green Solutions, an energy efficiency consultancy. "It's easier to reject heat to 60 degrees than it is to 100. And in the winter, it's a heck of a lot easier to absorb energy from 60 degrees than 32 degrees."

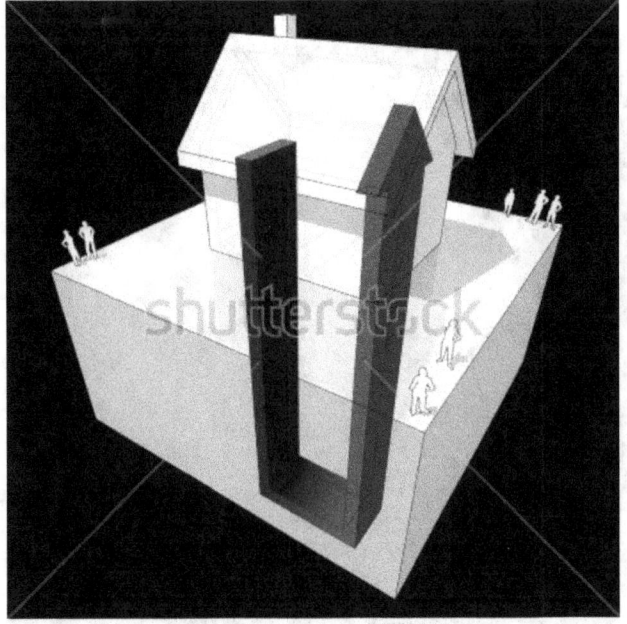

But with new home construction still in a slump and existing homeowners already looking to cut corners, intital costs are holding many homeowners back from taking the plunge into GSGHP.

"The first cost needs to be at the utility scale," said Ewbank. "Let the utilities install and finance them, then the homeowner could choose a very efficient system and not be subject to the first cost."

The idea is that utilities could then just bill their GSGE customers for BTUs of geo-cooling or heating rather than in kilowatt-hours.

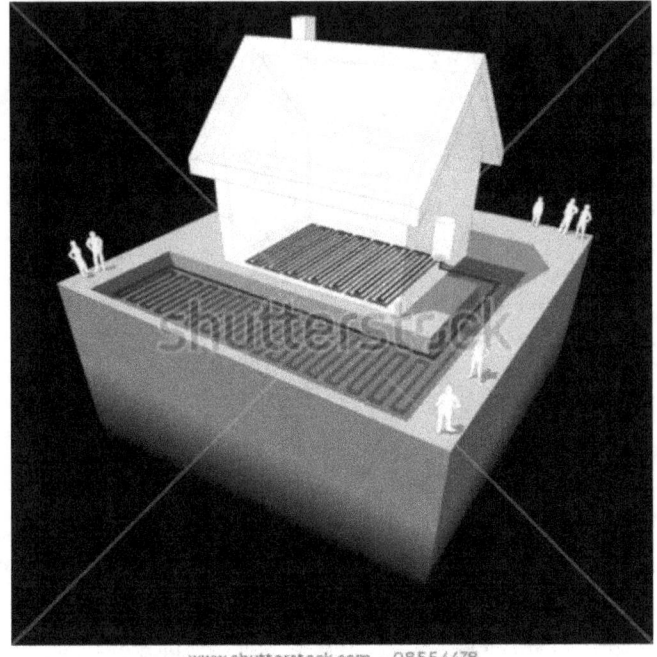

Ewbank says an electric utility would gain winter load since customers wouldn't shift from electricity to fossil fuels to heat their homes. And, in turn, gas utilities would gain summer load, since homes cooled by gas-driven heat pumps are the rare exception.

Ewbank says that, in essence, GSGHPs are no different than any other energy production facility. The difference is that competing energy technologies' higher "first costs" are usually borne by either the public at large or utilities involved in financing such projects.

Thus, Ewbank says, the challenge is to make sure that when future new homeowners argue over whether to forego Ground Source heat pumps in favor of granite kitchen countertops, they will understand that they have options to have both.

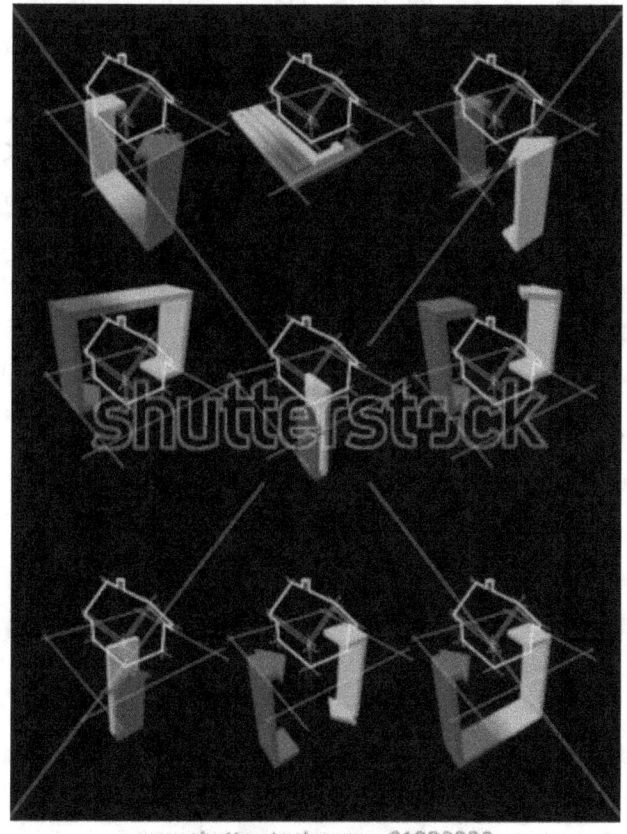

How can that happen? "We've got to shift the first costs of installation," said Ewbank.

In the meanwhile, Spitler says that newer, less expensive drilling techniques and improved heat exchangers would help make GSGE more attractive to both new homeowners and potential commercial users. But as Spitler points out, it's likely that the average U.S. homeowner, school or business doesn't realize that Ground Source Geothermal is a viable heating and cooling option.

However, Doug Dougherty, the president and CEO of the Geothermal Exchange Organization in Washington D.C., says that although new marketing efforts should help bolster GSGE's prospects, he doesn't foresee a significant increase in market-share until new home construction returns to pre-recession levels.

(From an article by Bruce Dorminey, in renewableenergyworld)

The idea of using Earth energy is old. Nikola Tesla always tried to get energy from the earth.

Cellars of houses which kept cool in summer and keep warm in the winter, were the easiest way to extract energy from the earth.

Wind farm

The cost of running and maintaining wind farms has fallen 38 percent in four years as competition among contractors increased and turbine performance improved, bringing closer the day that the technology matches fossil fuel.

The average price of operation and maintenance contracts for onshore farms this year slid to 19,200 euros ($25,000) a megawatt from 30,900 euros in 2008, Bloomberg New Energy Finance said today. It took data from 38 developers and service providers in more than 24 markets for its first O&M Price Index.

"Wind power has done much to improve its competitiveness against gas-fired and coal-fired generation in recent years via lower-cost, more technically advanced turbines and more sophisticated siting and management," Michael Liebreich, chief executive officer of London-based BNEF, said in the statement. The figures help inform debate in the U.S. and Britain, where governments are considering how much to subsidize renewable energy. BNEF's findings show manufacturers led by as Vestas Wind Systems A/S are bringing down prices, making the machines more profitable for power providers to run.

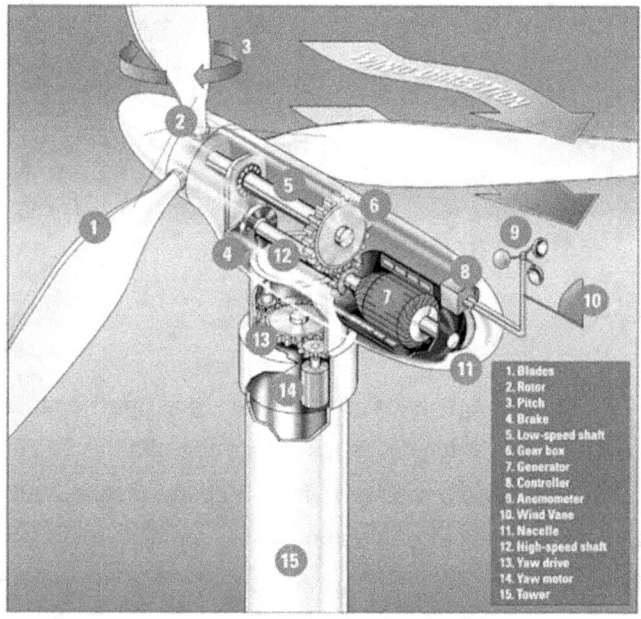

Britain is cutting subsidies for onshore wind by 10 percent as Prime Minister David Cameron's government discusses halting subsidies once the current renewable energy targets are met. In the U.S., President Barack

Obama's administration is pushing Congress to renew tax credits favoring wind in the face of opposition from Republican lawmakers.

Solar houses

Solar plants

Back in the beginning of the summer we heard about plans to develop what at the time would have been world's largest thin-film solar PV plant, a 10 megawatt facility outside of Las Vegas. Around the same time a 25 megawatt solar PV plant in Florida (using regular solar panels) was announced. While technically records, both really wouldn't provide that much power in the grand scheme of things.

Well, oh what a difference a couple of months can make. In the past month alone the scale of some of the new solar power plants being planned has increased such that you really should sit up at take notice. Granted, all of these are either in the planning stages or in the very first phases of construction—it wouldn't surprise if some of these plans get revised—but still, solar power plants that rival fossil fuel power plants in size is a huge boost for renewable energy.

250 MW Integrated Solar Power Facility Planned for West Bengal

The smallest and most recently announced of all these projects, Bhaskar Silicon Ltd will be building an integrated solar power complex in Haldia, West Bengal near the border with Bangladesh. The first phase of the project is expected to be completed by 2009. In addition to the 250 MW of electricity the facility will generate, by the time it is fully operational in 2011 it will be able to produce 5,000 tonnes of polysilicon.

BrightSource to Build 500 Megawatts of Solar-Thermal Power in Mojave Desert

Announced back in April, BrightSource Power will be building three solar thermal plants in the Mojave Desert of California with a combined capacity of 500 MW which are expected to come online by 2011. The utility buying the power will be PG&E;, which has signed contracts for an additional 400 MW of solar power which could bring the final size for this project to 900 MW.

550 Megawatts Thin Film + 250 MW Regular Solar for California

PG&E; apparently wasn't satisfied with its 500 MW arrangement with BrightSource above (hint: California's renewable targets have a lot to do with it...) and has signed contracts with two different solar developers using two different solar technologies: Optisolar will be building a record-shattering 550 MW thin-film solar plant in San Louis Obispo County, while SunPower will be building a 250 MW solar facility called the California Solar Ranch. The former project is expected to begin feeding power to the grid beginning in 2011, the latter in 2011.

World's Largest Solar Energy Project (5GW!) Planned for Gujarat, India

Though this project is by far the largest solar project under consideration, and would be one of the largest power projects regardless of source, the 5 GW (that's 5000 MW for the watt-conversion challenged) "Integrated Solar City" which was recently discussed by the Clinton Foundation and to be located in the western Indian state of Gujarat is such an increase in scale for solar power that, frankly, my jaw hangs open in a mixture of joy, disbelief and awe.

Even though it has not be disclosed whether the project will employ solar photovoltaic or solar thermal technology, it certainly is gigantic. Even if it eventually gets built at half its currently touted size, it'll still be bigger than your average nuclear plant by a wide margin.

9. Hydrogen Gas Production Doubled with New Super Bacterium

Hydrogen gas is today used primarily for manufacturing chemicals, but a bright future is predicted for it as a vehicle fuel in combination with fuel cells.

In order to produce hydrogen gas in a way that is climate neutral, bacteria are added to forestry or household waste, using a method similar to biogas production.

One problem with this production method is that hydrogen exchange is low, i.e. the raw materials generate little hydrogen gas. Now, for the first time, researchers have studied a newly discovered bacterium that produces twice as much hydrogen gas as the bacteria currently used.

The results show how, when and why the bacterium can perform its excellent work and increase the possibilities of competitive biological production of hydrogen gas.

(From http://www.alternative-energy-news.info/hydrogen-gas-production-doubled-new-super-bacterium/)

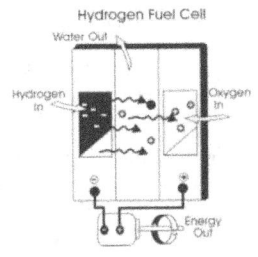

Splitting Water In to Hydrogen And Oxygen

We often want to imitate nature for near perfect results. But sometimes it just remains a desire. In its quest for green and clean energy mankind is searching for that magical method that can split water into hydrogen and oxygen. Nature performs this task wonderfully through the process of photosynthesis.

Man is still facing challenges in duplicating that process in the laboratory. If we are able to split water into oxygen and hydrogen in the presence of sunlight we will be able to harness the potential of hydrogen as a clean and green fuel. Till date man-made systems are quite inefficient, time consuming, money consuming and often require additional use of chemical agents.

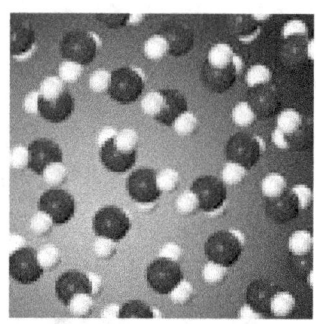

Researchers at the Weizmann Institute Organic Chemistry Department under the leadership of Prof. David Milstein have developed a novel way of splitting water molecules that can separate oxygen from water and bind the atoms in a different molecule. This technique leaves the hydrogen free to combine in other compounds as well. They were inspired by photosynthesis, a process carried out by plants. Photosynthesis is the life giving force on the earth because it is the source of all oxygen on the earth.

The new approach devised by the Weizmann team has three important steps that end in liberation of hydrogen and oxygen with the help of a special metal complex. This metal complex's core element is ruthenium. This 'smart' complex's metal part and organic part help in splitting the water molecules. When water is mixed with this complex, the bonds between the hydrogen and oxygen atoms break. Here one hydrogen atom binds with organic part of the complex, the hydrogen and oxygen atoms (OH group) bind to its metal center.

The second stage is known as heat stage. Here the water solution is heated up to 100 degrees C. This releases the hydrogen gas from the complex. Here comes our clean and green source of fuel. Another OH group is added to the metal center.

Milstein explains about the magical third stage, "But the most interesting part is the third light stage. When we exposed this third complex to light at room temperature, not only was oxygen gas produced, but the metal complex also reverted back to its original state, which could be recycled for use in further reactions."

The results are considered unique because of the generation of a bond between two oxygen atoms promoted by a man-made metal complex. It is a very unusual event. And it is still unanswerable how it can take place. The team has found out that during the third stage, light provides the energy for the two OH groups to get together to form hydrogen peroxide ($H2O2$). This hydrogen peroxide quickly breaks up into oxygen and water. What Milstein thinks about this chemical reaction? He says, "Because hydrogen peroxide is considered a relatively unstable molecule, scientists have always disregarded this step, deeming it implausible; but we have shown otherwise."

Another interesting thing that Milstein and his team has spotted is that the bond between the two oxygen atoms is generated within a single molecule. This bond formation doesn't occur between oxygen atoms located on separate molecules, but it comes from a single metal center.

The greatest achievement of Milstein's team has been the development of a mechanism for the formation of hydrogen and oxygen from water, without the need for sacrificial chemical agents. It has been achieved by using individual steps and utilizing light. For their next project, they intend to combine these stages to create a proficient catalytic system. These steps could leave a mark in the area of alternative energy.

(From http://www.alternative-energy-news.info/splitting-water-into-hydrogen-and-oxygen/)

Water into Hydrogen Fuel with Waste Energy

With each passing day, scientists are coming out with unique solutions to lessen our dependence on fossil fuels. They are now thinking of turning stray forms of energy such as noise or random vibrations from the environment into useful form of energy. They want to use piezoelectric effect for such purposes. Some materials produce electricity while undergoing mechanical stress. This is known as piezoelectric effect. Small piezoelectric crystals can come up with enough voltage to create a spark which can be utilized to ignite gas.

Piezoelectric crystals act as igniters. They are helpful in many gas-powered appliances like ovens, grillers, room heaters, and hot water heaters. These piezoelectric crystals are quite tiny and can be easily fitted into lighters too. Piezoelectric crystals are also fitted into electronic clocks and watches for time alarm noise.

Materials scientists at the University of Wisconsin-Madison have taken the help of piezoelectric effect to harness random energy available in the atmosphere to turn water into usable hydrogen fuel. It might prove a simple, efficient method to recycle waste energy. The research team is led by Huifang Xu, who is a UW-Madison geologist and crystal specialist. They took nanocrystals of zinc oxide and barium titanate. These two nanocrystals were put in water. When these crystals received ultrasonic vibrations, the nanofibers flexed and catalyzed a chemical reaction. This whole process resulted in splitting the water molecules into hydrogen and oxygen.

Huifang Xu along with his team has published their work in the Journal of Physical Chemistry Letters. They wrote in the journal, "This study provides a simple and cost-effective technology for direct water splitting that may generate hydrogen fuels by scavenging energy wastes such as noise or stray vibrations from the environment. This new discovery may have potential implications in solving the challenging energy and environmental issues that we are facing today and in the future."

Xu and his colleagues applied the piezoelectric effect to the nanocrystal fibers successfully. Xu says, "The bulk materials are brittle, but at the nanoscale they are flexible." It is akin to the difference between fiberglass and a pane of glass.

It has been noted that smaller fibers exhibit more flexibility than larger crystals. Therefore smaller fibers can generate electric charges without difficulty. The project team has extracted an impressive 18 percent efficiency with the nanocrystals, higher than most experimental energy sources. Xu shares his views, "because we can tune the fiber and plate sizes, we can use even small amounts of [mechanical] noise — like a vibration or water flowing — to bend the fibers and plates. With this kind of technology, we can scavenge energy waste and convert it into useful chemical energy." What a fantastic idea.

But scientists didn't utilize this electrical energy straightaway. They use this energy in breaking the chemical bonds in water to split oxygen and hydrogen. Xu explains, "This is a new phenomenon, converting mechanical energy directly to chemical energy." Xu calls it a piezoelectrochemical (PZEC) effect. Why it seems that scientists are beating around the bush? Because chemical energy of hydrogen fuel is more stable than the electric charge. Storage of hydrogen fuel is easy and would not lose potency over time.

With the right technology, Xu foresees this method to be utilized where small amount of power is needed. Now we can imagine charging a cell phone while taking our morning walk or we can enjoy cool breeze that can power street lights. Xu says, "We have limited areas to collect large energy differences, like a waterfall or a big dam. But we have lots of places with small energies. If we can harvest that energy, it would be tremendous."

New Hydrogen Storage Method

Hydrogen is an extremely environment friendly fuel as when it burns it releases only water vapor into the atmosphere, but the problem is that it is not easy to store it because it needs to be stored like other compressed gases. A new solid may solve the problem. A nonreactive noble gas called xenon combined with hydrogen and other massive pressure gives rise to a solid that can be later used to store hydrogen fuel. The research paper is published in the November 22, 2009, advanced online publication of Nature Chemistry. The discovery initiates a new line of materials that which could render impetus to new hydrogen technologies.

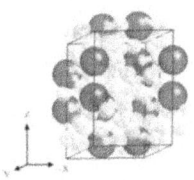

Xenon is used as an anesthesia, to preserve biological tissues, and it is also used in lighting. Being a noble, gas xenon does not typically react with other elements. Researchers used a diamond anvil device to squeeze together xenon and hydrogen.

The lead author Maddury Somayazulu, research scientist at Carnegie's Geophysical Laboratory, explained: "Elements change their configuration when placed under pressure, sort of like passengers readjusting themselves as the elevator becomes full. We subjected a series of gas mixtures of xenon in combination with hydrogen to high pressures in a diamond anvil cell. At about 41,000 times the pressure at sea level (1 atmosphere), the atoms became arranged in a lattice structure dominated by hydrogen, but interspersed with layers of loosely bonded xenon pairs.

101

When we increased pressure, like tuning a radio, the distances between the xenon pairs changed-the distances contracted to those observed in dense metallic xenon."

The researchers imaged the compound at different pressures using X-ray diffraction, infrared and Raman spectroscopy. They were really surprised to realize that the response of xenon with the surrounding hydrogen was responsible for the unusual stability and the continuous change in xenon-xenon distances as pressure was adjusted from 41,000 to 255,000 atmospheres.

The researchers were taken off guard by both the structure and stability of this material xenon, according to the lead crystallographer who looked at the changes in electron density at varying pressures using single-crystal diffraction. As electron density from the xenon atoms spreads towards the surrounding hydrogen molecules, it stabilizes the compound and the xenon pairs.

Xenon as a hydrogen carrier is too heavy and expensive but further research in this direction can definitely lead to lighter alternatives.

According to Russell Hemley, director of the Geophysical Laboratory and a co-author, "This hydrogen-rich solid represents a new pathway to forming novel hydrogen storage compounds and the new pressure-induced chemistry opens the possibility of synthesizing new energetic materials."

(From http://www.alternative-energy-news.info/new-hydrogen-storage-method/)

Saving energy can become a new energy source

In our energy deprived world scientists are trying to find out various elements, alloys and substances that can provide clean and green energy along with meeting our energy demands. This quest has led them to superconductors.

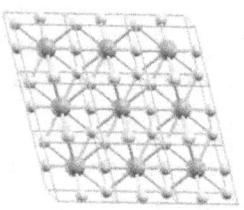

Superconductor materials have no electrical resistance. This property paves way for electrons to travel through them freely. Superconductor materials also carry large amounts of electrical current for long periods of time without losing energy as heat. Scientists are of the view that metallic hydrogen can prove to be a high-temperature superconductor.

We know that hydrogen is the lightest element present in this universe and lithium is the third lightest element. We find hydrogen as gas and lithium as metal at given temperature and pressure present on the earth. In hydrogen gas, the atoms are robustly paired and each hydrogen atom contributes one electron to the bonding. In chemistry shorthand, hydrogen is called H2.

Proceedings of the National Academy of Sciences published a paper this week written by a team of scientists from Cornell University and the State University of New York. They shared that if we add small amounts of lithium to hydrogen and if we keep the pressure at about one-fourth of a pressure (on which hydrogen turns into a metal) hydrogen transforms into a metal with superconductivity properties. National Science Foundation (NSF) has provided finances for this project.

Hydrogen and lithium react with each other and form a non-metallic stable compound. This is shown as LiH and known as lithium-hydrogen compound. Jupiter and Saturn experience intense gravitational forces and pressures therefore hydrogen is found in metallic form there.

Scientists are trying to create situations so that they can extract hydrogen's electron. How? They are trying to squeeze it between the facets of a diamond anvil cell under pressures up to 3.4 million atmospheres. We know that the atmospheric pressure at sea level is one atmosphere and at the center of the earth is around 3.5 million atmospheres. Scientists have faced lots of difficulties with this method of steady pressures. They have been trying shock-wave methods. This is a kind of sophisticated computer program.

The programs theoretically compute if hydrogen can be metalized by uniting a lithium atom with varying numbers of hydrogen atoms. The programs also calculate if metallic hydrogen can be made under pressures attainable in a laboratory.

The lithium and hydrogen combinations predicted by the study have not been verified in a laboratory till now.

The research team is trying out various combinations of hydrogen and lithium. One of the combinations contains one lithium atom for every six hydrogen atoms or LiH6. The complex calculations forecast that in the imaginary compound the Li atom is activated to discharge its solitary outer electron, which is then distributed over the three H2 molecules. It has already been confirmed that under pressure, the hypothetical reaction forms a stable and metallic hydrogen compound.

The calculations also foresee that LiH6 could be a metal at normal pressures. However, under these conditions it is unstable and would decompose to form LiH and H2.

"The stable and metallic LiH6 compound is predicted to form around 1 million atmospheres, which is around 25 percent of the pressure required to metalize hydrogen by itself," said Eva Zurek, lead author of the paper and an assistant professor of chemistry at The State University of New York, Buffalo.

"Interestingly, between approximately 1 and 1.6 million atmospheres, all the LiH combinations studied were stable or metastable and all were metallic," said

Roald Hoffmann, co-author, recipient of the 1981 Nobel Prize in chemistry and Cornell's Frank H.T. Rhodes Professor of Humane Letters, Emeritus.

Another one of the hypothetical compounds studied by the team was composed of one lithium atom and two hydrogen atoms or LiH_2 (see bottom right image).

"The theoretical study opens the exciting possibility that non-traditional combinations of light elements under high pressure can produce metallic hydrogen under experimentally accessible pressures and lead to the discovery of new materials and new states of matter," said Daryl Hess, a program director in the NSF Division of Materials Research.

"Once again, these researchers have taken chemistry to a new frontier," said Carol Bessel, a program director in the NSF Division of Chemistry. "They have described, through their theories and calculations, molecules that test our fundamental assumptions about atoms, molecules and structures. In doing so, they challenge the experimentalists to make what they have imagined in their minds a reality to be held in the hand."

The team members believe the information gleaned from the study suggests that one may combine large amounts of hydrogen with other elements. The information may also someday assist with the design of a metallic hydrogen-based superconductor.

Neil W. Ashcroft, who is the co-author, and Cornell's Horace White Professor of Physics, Emeritus says, "We have already been in touch with laboratory experimentalists about how LiH6 might be fabricated, starting perhaps with very finely divided forms of the common LiH compound along with extra hydrogen."

(From http://www.alternative-energy-news.info/lithium-hydgrogen-production/)

Hydrogen-Producing Bacteria Provide Clean Energy

Science Daily (Aug. 26, 2008) — A new "green" technology developed cooperatively by scientists with the Agricultural Research Service (ARS) and North Carolina State University (NC State) could lead to production of hydrogen from nitrogen-fixing bacteria.

Renewable sources of energy—such as hydrogen— that don't produce pollutants or greenhouse gases are needed to solve global energy shortages. Fossil fuels such as coal, oil and natural gas are nonrenewable energy sources implicated in global warming.

The invention holds promise as a source of hydrogen for use in fuel cell technology. Fuel cell devices combine hydrogen and oxygen to produce electricity and water, and are considered efficient, quiet and pollution-free.

Fuel cells are now being tested in a range of products, including automobiles that release no emissions other than water vapor.

ARS inventors Paul Bishop and Telisa Loveless and NC State inventors Jonathan Olson and José Bruno-Bárcena developed the patent-pending technology.

Nitrogen-fixing bacteria play a key role in agriculture. They live in soil and on certain plant roots, and convert nitrogen from the air into a chemical form that plants can use to grow. The researchers developed a way to identify strains of these bacteria that produce hydrogen gas.

Bishop first demonstrated novel aspects of bacterial nitrogen-fixing more than two decades ago. Building on that work, the team developed a method that uses a selecting agent to identify these special hydrogen-producing strains. The selecting agent allows researchers to identify these bacterial strains without the need for genomic sequencing or genetic modification.

Using the selecting agent, the inventors identified a gene that inactivates the bacteria's hydrogen uptake system so that all of the hydrogen produced is released. Because the bacterial cells cannot recycle the hydrogen, the hydrogen they produce can be captured and used as a fuel whose byproduct is water and heat.

Licensing information can be obtained by contacting the ARS Office of Technology Transfer or the Office of Technology Transfer at NC State.

ARS is a scientific research agency of the U.S. Department of Agriculture

(From http://www.sciencedaily.com/releases/2008/08/080825195852.htm)

10. Capturing energy concentrated near the source and forwarding

directly to Earth in concentrated form

Should start some spatial projects, to capture a large amount of energy somewhere near the source (near the Sun), energy which can be sent then to the Earth in a concentrated form (LASER, MASER, IRASER, etc).

The enormous energy emanating from the sun is spreading in all directions of the universe, and dilute with the distance.

On Earth no longer reach than a small amount from the energy emanated by the sun.

We try here (on the Earth) to capture a drop from a very small amount of energy, who came from Sun. And we also complain that the yield is low, and technological costs are high.

In the next figure we can see how a large amount of energy is transmitted to long distances with low losses, naturally, because is emitted by a sun (a star) in concentrated form, with natural lasers.

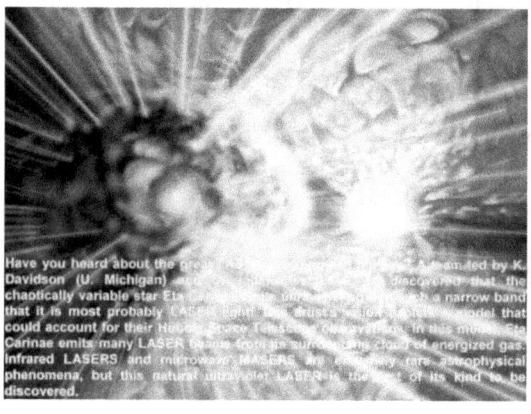

This is exactly what should we do. This sun strange and extremely rare in Universe, shows us what must we do.

In the next figure we can see the exact position of our planet in our solar system.

It can see as well how the sun's energy is diluted when the distance from sun grows.

The third halo surrounds the planets Mercury and Venus, and barely touching the Earth.

The fourth halo (the most pale from those which are visible with the naked eye) reach Jupiter.

Mercury is hot, and Saturn is cold.

Installations which must do capturing the solar energy, could be installed over the Mercury. From the Mercury, the concentrated energy will be transmitted directly focused on the Moon.

On the Moon, the energy will be conserved and forwarded to Earth in doses non-hazardous (with lower concentrations), using multi-channels microwaves.

Final Conclusions

Renewable energy is energy which comes from natural resources such as sunlight, wind, rain, tides, and geothermal heat, which are renewable (naturally replenished). In 2008, about 19% of global final energy consumption came from renewables, with 13% coming from traditional biomass, which is mainly used for heating, and 3.2% from hydroelectricity.

New renewables (small hydro, modern biomass, wind, solar, geothermal, and biofuels) accounted for another 2.7% and are growing very rapidly. The share of renewables in electricity generation is around 18%, with 15% of global electricity coming from hydroelectricity and 3% from new renewables.

This book aims to disseminate new methods of obtaining energy.

After 1950, began to appear nuclear fission plants. The fission energy was a necessary evil. In this mode it stretched the oil life, avoiding an energy crisis. Even so, the energy obtained from oil represents about 66% of all energy used.

At this rate of use of oil, it will be consumed in about 40 years.

Today, the production of energy obtained by nuclear fusion is not yet perfect prepared.

But time passes quickly. We must rush to implement of the additional sources of energy already known, but and find new energy sources.

In these circumstances this book comes to proposing possible new energy sources.

Bibliography

[1] EWEA Executive summary "Analysis of Wind Energy in the EU-25" (PDF). European Wind Energy Association. http://www.ewea.org/fileadmin/ewea_documents/documents/publications/WETF/Facts_Summary.pdf EWEA Executive summary. Retrieved 2007-03-11.

[2] Massachusetts Institute of Technology (2010, September 13). Funneling solar energy: Antenna made of carbon nanotubes could make photovoltaic cells more efficient. *Science Daily*. Retrieved September 21, 2010, from http://www.sciencedaily.com/releases/2010/09/100912151548.htm

[3] "Towards Sustainable Production and Use of Resources: Assessing Biofuels". United Nations Environment Program. 2009-10-16. http://www.unep.fr/scp/rpanel/pdf/Assessing_Biofuels_Full_Report.pdf. Retrieved 2009-10-24.